I0797975

MY FIRST BOOK OF
EVOLUTION

SHEDDAD KAID-SALAH FERRÓN
& EDUARD ALTARRIBA

CONTENTS

LIFE

LIFE IS EVERYWHERE ON OUR PLANET

Wherever you look, you can find animals, plants, and microbes in millions of shapes, sizes, and colors. And the most amazing thing is that all these living beings are adapted to their surroundings. It's as if they have been deliberately designed for the habitats where they live.

Bats use radar to find and hunt insects.

The blue whale's huge mouth filters up to 90 tonnes of water to catch krill, the tiny marine creatures it feeds on.

Flowers are pollinated when a hummingbird flies from one flower to another.

The hummingbird can hover in the air while it sucks the nectar from a flower with its long beak.

Fish use their gills to breathe underwater. They have fins and a tail for swimming and are the ideal shape for moving through water.

There is an impressive variety of living things on Earth, such as plants and animals. But how can we explain this rich diversity when we know that living things don't appear all of a sudden and aren't created out of thin air? How can we explain why they all seem to be perfectly designed to live how they live and where they live?

This puzzle was solved in the mid-19th century, mostly thanks to the work of the English scientist and naturalist Charles Darwin (1809–1882), who put forward the idea of evolution through natural selection. Darwin realized that biodiversity is mainly due to the variations that different species undergo in order to adapt to the different environments that have existed on Earth throughout time.

THE FAMILY TREE

Your closest ancestors are your parents: your mother and your father. You and your cousins share older ancestors: your grandparents. Going a bit farther back, you, your mother, your aunts and uncles, your cousins, and your grandmother share a more distant relative: your great-grandmother. We can show all your family relations in the form of a family tree.

In the branches closest to the bottom of the tree trunk are the youngest members of the family. As we go further up the trunk, we see older people who were born longer ago.

If we keep going up, we'll find family members who are even older or perhaps no longer with us. They are the ancestors of everyone below them.

One of the basic ideas of biology is that all living things are related: we are a family. And if we're a family, we can have a family tree, can't we?

THE TREE OF LIFE

The tree of life is the family tree of all living things that have ever lived on Earth.

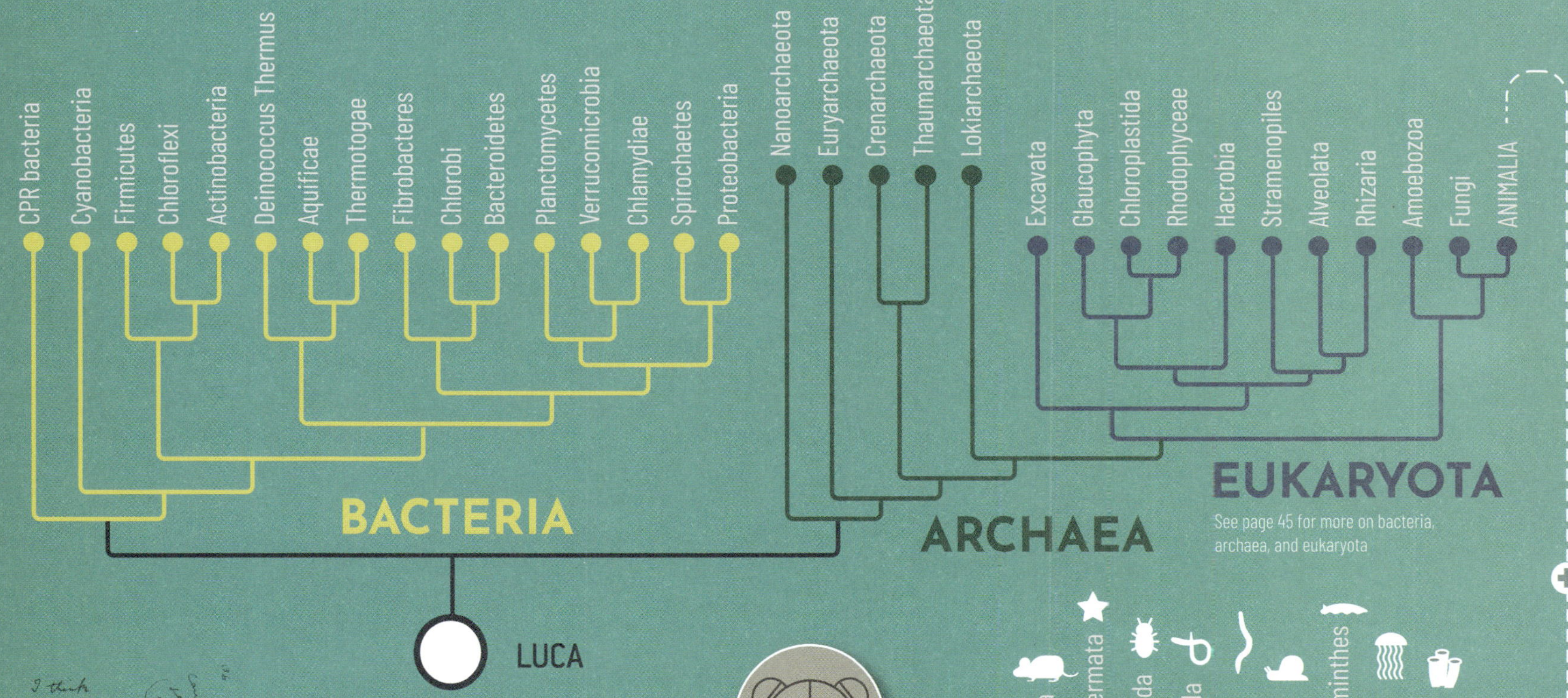

CPR bacteria
Cyanobacteria
Firmicutes
Chloroflexi
Actinobacteria
Deinococcus Thermus
Aquificae
Thermotogae
Fibrobacteres
Chlorobi
Bacteroidetes
Planctomycetes
Verrucomicrobia
Chlamydiae
Spirochaetes
Proteobacteria

Nanoarchaeota
Euryarchaeota
Crenarchaeota
Thaumarchaeota
Lokiarchaeota

Excavata
Glaucophyta
Chloroplastida
Rhodophyceae
Hacrobia
Stramenopiles
Alveolata
Rhizaria
Amoebozoa
Fungi
ANIMALIA

BACTERIA

ARCHAEA

EUKARYOTA

See page 45 for more on bacteria, archaea, and eukaryota

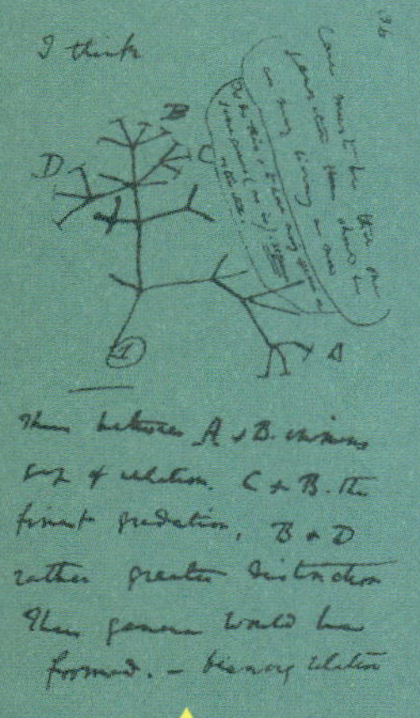

LUCA

Charles Darwin (see page 14) drew a phylogenetic tree in his notebook.

Our first common ancestor, the LUCA (see page 44) sits at the base of the tree trunk. Extending outward are branches representing all the different organisms that have lived on Earth, right up to today's diverse life-forms.

Trees like these that represent evolution are called PHYLOGENETIC TREES.

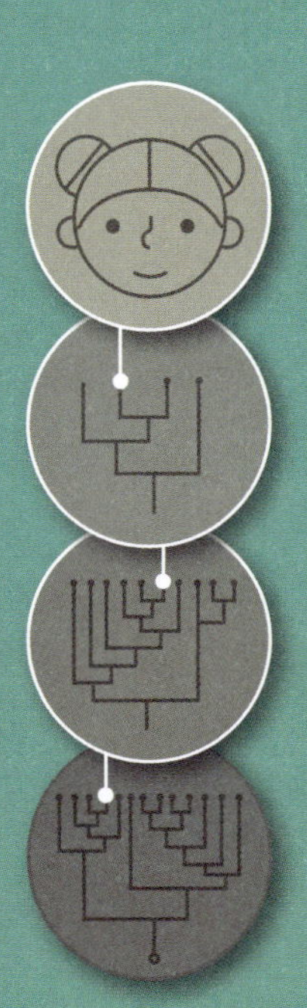

The tree of life has so many branches that they cannot be shown in a single drawing. So, we can draw smaller branches to see the living things that are closely related to each other.

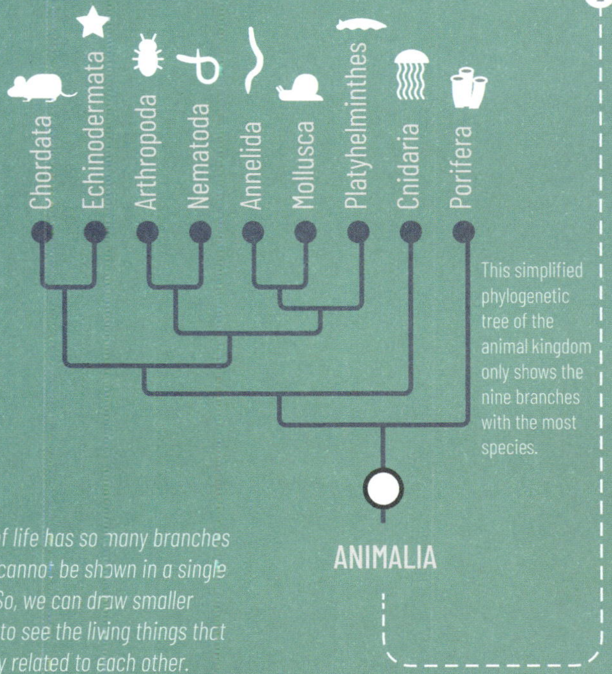

Chordata
Echinodermata
Arthropoda
Nematoda
Annelida
Mollusca
Platyhelminthes
Cnidaria
Porifera

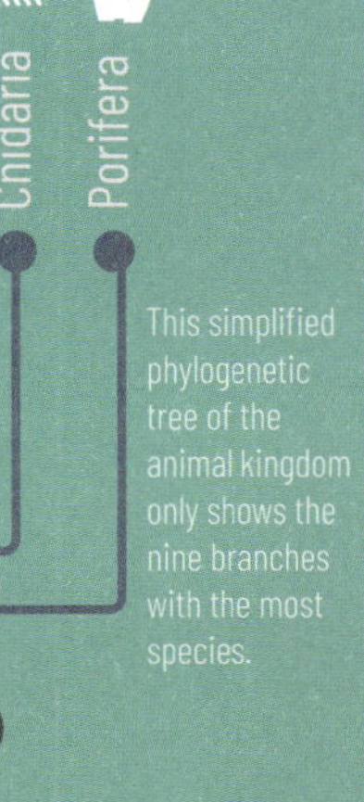

This simplified phylogenetic tree of the animal kingdom only shows the nine branches with the most species.

ANIMALIA

We're going to talk about species a lot in this book. We already know more or less what that means. For example, you know that cats and dogs aren't the same species, nor are gorillas and chimpanzees, nor are roses and cucumbers. We think of each of these groups of living things as different species.

But what exactly is a species?

We can define a *species* as a group in which all members share **common characteristics** and **can reproduce** and exchange genes among each other, meaning that they produce **descendants** with the same characteristics as their parents and, very importantly, are **fertile.**

For example, we say that **horses are a species**. If two horses reproduce, their children (their offspring) will also be horses that grow up, reproduce, and have more baby horses.

Donkeys (and asses) belong to a **different species**. When they reproduce, their offspring will have typical donkey characteristics and will be able to go on to have baby donkeys.

These two species are so closely related that they can reproduce with each other. When a mare (a female horse) and a male donkey mate, their offspring is **a mule, a hybrid of the two species** with donkey characteristics (such as narrow hoofs and long ears) and horse characteristics (such as size and height).

However, **mules are sterile** and cannot reproduce. That's how we know that horses and donkeys, although very similar, are different species.

Although our definition of species makes everything seem very clear, in actual fact it isn't: there are many cases in nature where this definition doesn't really apply. For example, many bacteria reproduce without needing to mate with another bacterium: when they reach a certain size, each bacterium simply divides into two clone daughter bacteria.

The idea of species is something we humans use to classify living organisms. But we have to keep in mind that things in nature are not always what we think they are.

How organisms look on the outside isn't the only way to determine how similar they are. These days scientists can also compare the DNA inside their cells (see page 23).

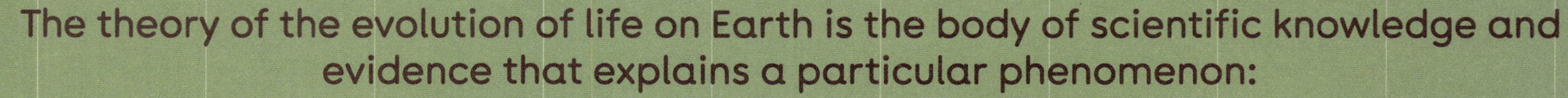
The theory of the evolution of life on Earth is the body of scientific knowledge and evidence that explains a particular phenomenon:

EVOLUTION

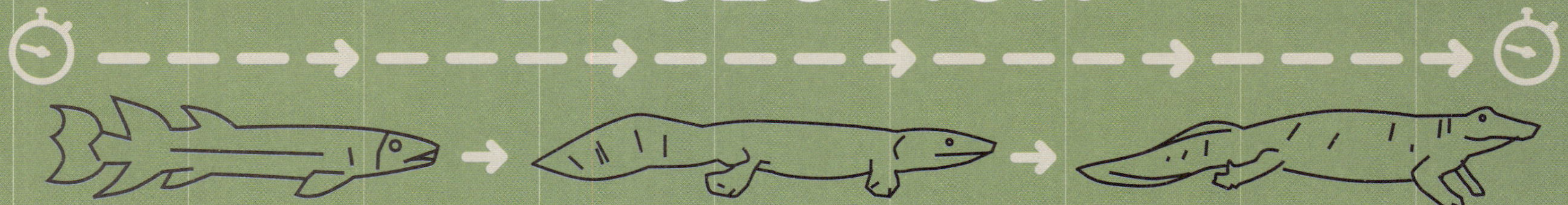

When we say that something evolves, what we are saying is that it changes and transforms over time.

Evolution is the natural process of species adapting, through changes that take place in individuals over generations. It has shaped all forms of life.

These changes are due to genetic mutations, which are errors that happen when copies of DNA are made (see page 24) and passed on from parents to children.

Palaeotragus

Miocene

Bramatherium

Late Miocene

Ardipithecus
around 4.4 million years ago

Australopithecus
around 2 million years ago

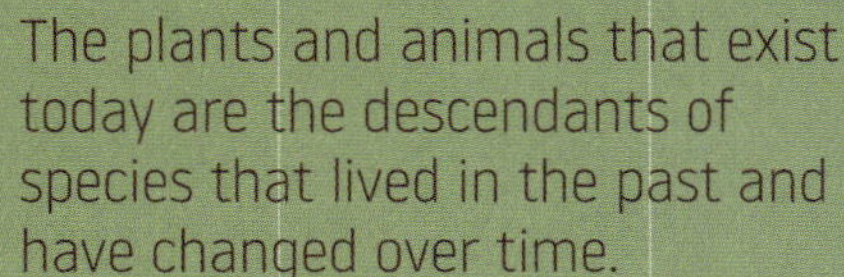

The plants and animals that exist today are the descendants of species that lived in the past and have changed over time.

Evolution is usually an extremely slow process in which small changes are inherited and build up from one generation to the next over millions of years.

Some animals, such as ants and sharks, have barely changed at all over millions of years. Their basic design is so well adapted to their environment that the process of evolution has barely changed their structure and characteristics.

Giraffe

Plants appeared around 140 million years ago. Their ancestors were plants that produced seeds but didn't bear fruit, like pine and fir trees nowadays.

Giraffes have changed from one generation to the next over a very long time. This has given them the characteristic appearance they have now, with their slender legs and long necks.

The same happened with us humans: we evolved from primates that lived millions of years ago and were very different from us.

CAREFUL! NOT ALL CHANGES ARE EXAMPLES OF EVOLUTION

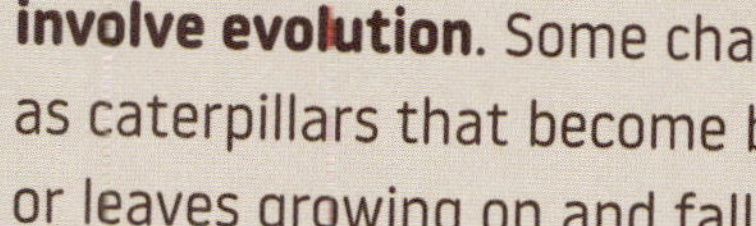

Not all changes that happen over time **involve evolution**. Some changes, such as caterpillars that become butterflies, or leaves growing on and falling from trees, are part of the natural life cycle, while receiving a life-changing injury is an accident. None of these will change what the descendants inherit.

These kind of changes that happen to living things throughout their lives are NOT evolution.

Now

Homo sapiens

NATURAL SELECTION

Life is not easy. All creatures must search for food, protect themselves from the cold and the heat, breathe, and avoid danger; in short, they must survive so they can leave descendants.

Living things are very different from one another, including within the same species. We have our own traits and characteristics that make us unique and distinguish us from the rest. There can be many kinds of differences, some very obvious, some not so obvious.

The great English naturalist Charles Darwin (see page 14) noticed that the traits an individual is born with give them a better or worse chance of survival.

Individuals that are better adapted to their environment survive longer and have more chance of producing descendants, to whom they can pass on these helpful traits. So, as time goes on, a larger number of individuals that are better adapted to their surroundings will be born and the species will evolve.

Darwin used this idea of natural selection to explain evolution in his famous book *The Origin of Species*.

On the previous page, we saw how today's giraffes evolved over millions of years. Now we're going to see how natural selection helped to bring about these changes.

Different traits

In a giraffe population, not all individuals have the same length necks or legs. Some will be taller than others.

INDIVIDUALS DON'T EVOLVE, THE POPULATIONS IN A SPECIES DO.

There cannot be an unlimited number of species in a particular environment.

Each species makes the most of its adaptive advantage.

Giraffes feed on tall plants. They eat leaves from trees and shrubs above ground level that other animals can't reach.

RESOURCES ARE LIMITED

Because there aren't an infinite number of trees, the environment can't support an infinite number of giraffes or other herbivores.

The tallest giraffes can reach the higher-up leaves but the shortest giraffes can't. When all the lower leaves are gone, giraffes with shorter necks or legs have less chance of survival.

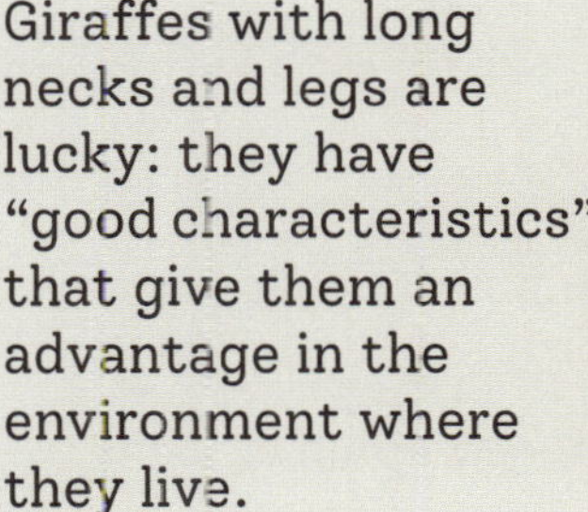

Giraffes with long necks and legs are lucky: they have "good characteristics" that give them an advantage in the environment where they live.

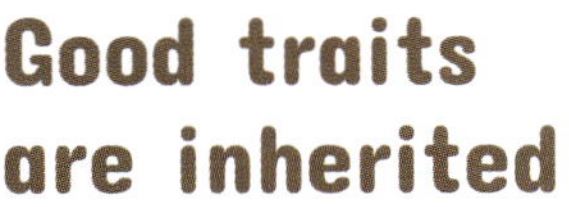

Good traits are inherited

Thanks to their traits, tall giraffes are more likely to survive, reproduce, and pass on their "good characteristics" to their offspring, which inherit long necks and legs.

Final result

Over time, the most useful traits—those that allow giraffes to have more descendants—become more common in the population and there are greater numbers of taller giraffes.

The voyage of the BEAGLE

Charles Darwin

1809–1882

Charles Darwin was an English naturalist. Even as a child, he was fascinated by nature and collecting shells, minerals, and other objects.

As the son and grandson of physicians, Darwin went to the University of Edinburgh to study medicine. But he found his studies so boring and unpleasant that his father sent him to Cambridge to study to become a clergyman. There he attended botany, entomology (insects), and geology classes, and met some important naturalists. One of them, John Henslow, became his mentor and friend.

It was Henslow who recommended Darwin for the job of ship's naturalist aboard **HMS *Beagle***. It was a voyage that would change his life.

For almost **five years, Darwin traveled around the world**, studying, collecting, and investigating biology, geology, paleontology, and other disciplines. His observations got him **thinking about species and evolution**.

After his voyage, **Darwin carried on researching the evolution of species for many years**. His conclusions and theories caused huge uproar and controversy, but today they remain fundamental to modern biology.

N

Galápagos Islands
SEPTEMBER 1835
One of the most important stops in the history of science.

Lima, Peru
JULY 1835

New Zealand
DECEMBER 1835

Valparaíso, Chile
JULY 1834
On an expedition to the Andes, Darwin explored and studied the region's geology and biodiversity.

Strait of Magellan and Cockburn Channel
JUNE 1834

The mission of HMS *Beagle* and its captain, Robert FitzRoy, was to chart the dangerous southern coastline of South America and to carry out calculations around the globe to determine longitude.

FitzRoy was only twenty-six years old at the time. He had taken command of the ship three years earlier, following the death of the previous captain on the *Beagle*'s first scientific voyage to South America.

The ship carried the latest technology to make measurements. This included instruments such as chronometers (used to determine the ship's exact position). The cannons were made of copper because the usual iron ones would have interfered with the compasses and chronometers.
Plymouth, England
DECEMBER 27, 1831 and OCTOBER 2, 1836
The Beagle set sail on December 27, 1831 and didn't return to England for almost five years, landing at Falmouth.
e Azores, Portugal
PTEMBER 1836
Canary Islands
The crew couldn't disembark in Tenerife because of a cholera quarantine.
Cape Verde
JANUARY 1832
Darwin studied volcanoes and observed octopuses, noting that they produced different colored ink depending on their hiding place.
Ascension Island
JULY 1836
St Helena
JULY 1836
Rio de Janeiro, Brazil
APRIL 1832
Montevideo, Uruguay
JULY 1832
Argentine coast and Tierra del Fuego
From July 1832 to June 1834, the Beagle mapped the Argentine coast. There were several missions inland, on which Darwin found different fossils of extinct mammals.
Falkland Islands
MARCH 1833
Cape Horn
DECEMBER 1833—JANUARY 1834
Cape Town, South Africa
MAY 1836
Darwin visited John Herschel, an English mathematician and astronomer who provided inspiration for The Origin of Species.
Port Louis, Mauritius
APRIL 1832
Darwin studied the coral reefs and came up with a theory of how coral atolls were formed.
Cocos (Keeling) Islands
APRIL 1836
Studying platypuses and the differences between marsupials in each region helped Darwin develop his theories further.
Sydney, Australia
JANUARY 1836
Albany, Australia
MARCH 1836
Tasmania, Australia
FEBRUARY 1836
The Beagle belonged to the British Royal Navy. It was only 90 feet (27.5 meters) long and carried 74 passengers and crew.
The Beagle was a boat full of boats. Whenever it reached a destination, it launched its fleet of seven small boats so the scientists could do their work.

DEVELOPING THE THEORY OF NATURAL SELECTION

During his voyage, Darwin observed a wide diversity of species and noticed that sometimes **individuals in the same family of animals showed differences** in appearance, size, and behavior that seemed to be related to their environment.

In the Galápagos Islands, Darwin observed finches and saw that these birds varied in size and had different shaped beaks from one island to another. He already knew of the theories of the geologist **Charles Lyell**, who spoke about gradual changes to the Earth taking place over very long periods of time.

He thought that the same thing could happen with species. Following his intuition, he came to the conclusion that the **finches on the Galápagos Islands** might have descended from a common ancestor and adapted to different environments over a long period of time: FINCHES **HAD CHANGED**! (See page 32.)

Geospiza magnirostris

Geospiza parvula

Certhidea olivacea

Geospiza fortis

On his journey back to England, Darwin studied his observations in depth and assembled a huge amount of information on variation and natural selection in nature.

He also read the work of Thomas Malthus, who argued that the human population was growing faster than the resources needed to support it, and that this would lead to a struggle for survival.

"The Origin of Species"

By the time the *Beagle* returned to England on October 2, 1836, Darwin had become a celebrity. He wrote many books on many subjects, including *The Voyage of the Beagle* and *The Descent of Man and Selection in Relation to Sex*. But his most important work is undoubtedly *The Origin of Species* (1859), in which he presents his theory of evolution by natural selection.

Alfred Wallace

Darwin had spent years developing his theories on natural selection and evolution when he received a manuscript from a naturalist called Alfred Wallace. It contained the same ideas, based on Wallace's studies of the Malay Archipelago. Both men were astonished and decided to present their theories together at the Linnean Society of London. There Wallace's manuscript was read out alongside an extract from Darwin's theories that later became *The Origin of Species*. This is why we often refer to the theory of evolution as the "Darwin-Wallace theory."

THE ORIGIN OF SPECIES **IS ONE OF THE MOST INFLUENTIAL SCIENCE BOOKS IN HISTORY.**

The ideas and concepts in the book challenged some of the religious beliefs of society at the time, especially creationism, which held that everything (humans, plants, animals, and the world as we know it) had been created by God.

ADAPTATION TO THE ENVIRONMENT

The harmless Sinaloan milk snake (*Lampropeltis triangulum sinaloae*) is a similar color to the coral snake, a very venomous species. With this disguise, predators don't dare come too close.

Some microorganisms have adapted to an environment with hardly any oxygen. They live in underwater vents, places where there are gas and lava eruptions.

The impala (*Aepyceros melampus*) is an antelope known for its giant leaps that can cover distances of over 32 feet (10 meters). It uses this ability to escape its many predators, such as leopards, cheetahs, African wild dogs, lions, hyenas, crocodiles, and pythons.

The tiger (*Panthera tigris*) uses its striped coat to camouflage itself in grass and get close to its prey without being seen.

The rose of Jericho (*Anastatica hierochuntica*) is a plant found in deserts in the Middle East that can completely dry out and stay curled up into a ball for years. When humidity increases, it turns green and opens up again.

Moles (*Talpidae*) live underground and have very poor eyesight and hearing. Their front claws are designed to dig out the burrows they live in.

The monarch butterfly (*Danaus plexippus*) travels from Mexico to Canada, covering around 3,000 miles (4,800 kilometers). During its two-month journey, several generations of butterflies are born in different places. Later, they return to the tree their ancestors departed from.

Barn owls (*Tyto alba*) have developed night vision to be able to hunt (and avoid being hunted) at night.

HEREDITY

It is no secret that children tend to look like their parents; some of the parents' characteristics are passed on to their children. This is true of all living things. If two dogs have a litter, the puppies look like their parents and also look like each other (some more so, others less so). Obviously, you won't find a mastiff, let alone a duck, in a litter of Chihuahuas.

In other words, individuals have characteristics that can be inherited by their descendants.

But how these inheritable traits are passed on has always been a mystery. No one knows for sure. Even Darwin admitted: "The laws governing inheritance are for the most part unknown."

The key is the genes:

The first person to give a thorough explanation of genetic inheritance was **Gregor Mendel** (1822–1884).

Mendel loved plants and knew a lot about them. He used pea plants for his scientific experiments because he knew they were easy to work with and crossbreed to produce daughter plants.

Pea plants have a number of easily recognizable inheritable characters:

Peas can be green or yellow in color.

Peas can be rough or smooth.

The flowers can be purple or white.

We call each of these inheritable characters a gene.

color gene | **roughness gene** | **flower gene**

The offspring inherit two versions of the gene, one from the mother and the other from the father.

Each version of a gene is called an **allele**.

In his experiments crossbreeding plants, Mendel noticed that each pea plant parent could pass on these characters or genes to their descendants. So each daughter plant receives one allele from the mother and one allele from the father.

MOTHER PLANT — ALLELE from the MOTHER

FATHER PLANT — ALLELE from the FATHER

DAUGHTER PLANT — RECESSIVE ALLELE: ALLELE from the FATHER; DOMINANT ALLELE: ALLELE from the MOTHER

*For example: we take a mother plant that has **yellow** peas and we cross it with a father plant that has **green** peas to make daughter plants.*

*Each daughter plant receives a **yellow** allele from the mother and a **green** allele from the father.*

*But when we look, the peas on the **daughter plant are yellow** and not green.*

*This means that the yellow color dominates the green color. In this case, the **yellow** allele, from the mother, is the one that is expressed and that we can see. We call it a **dominant allele**.*

*Whereas the father's **green** allele remains hidden and is not expressed, so we call it a **recessive allele**.*

And if we cross two daughter plants, we could get green peas, because one of the daughter plants could get the green allele from both the mother and father. The grandfather plant's green color has not been lost and has reappeared.

In the same way, roughness and flower color genes are passed on through their own alleles.

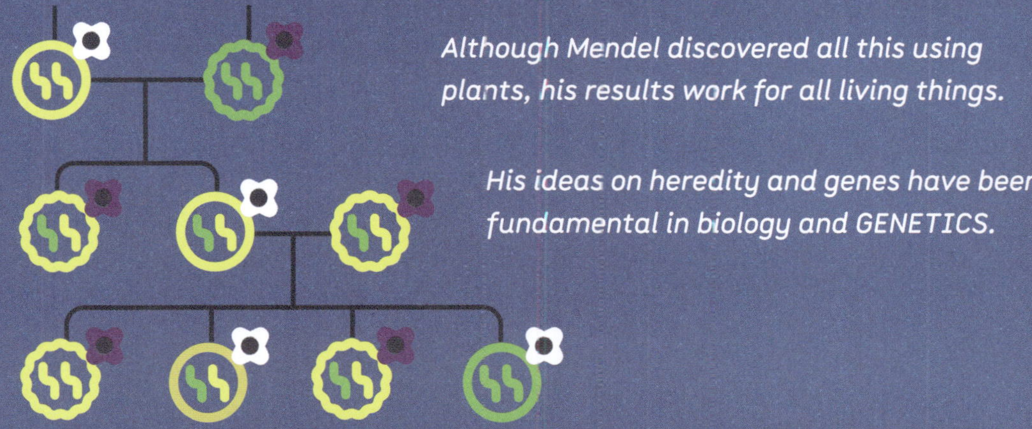

Although Mendel discovered all this using plants, his results work for all living things.

His ideas on heredity and genes have been fundamental in biology and GENETICS.

CELLS

A cell is the basic unit of life.

It is the smallest and most basic structure that can carry out all the functions needed for life. All living organisms are made up of cells, from bacteria (made from a single prokaryotic cell) to animals, fungi, and plants (made from many eukaryotic cells, which are more complex).

Groups of cells form the tissues of multicellular animals: muscles, brain, cartilage, and so on.

Cells have a complex and organized structure that allows them to carry out all the functions needed for life. They are surrounded by a cell membrane and inside is the genetic material (DNA). Cells also have specialized parts called organelles for specific functions.

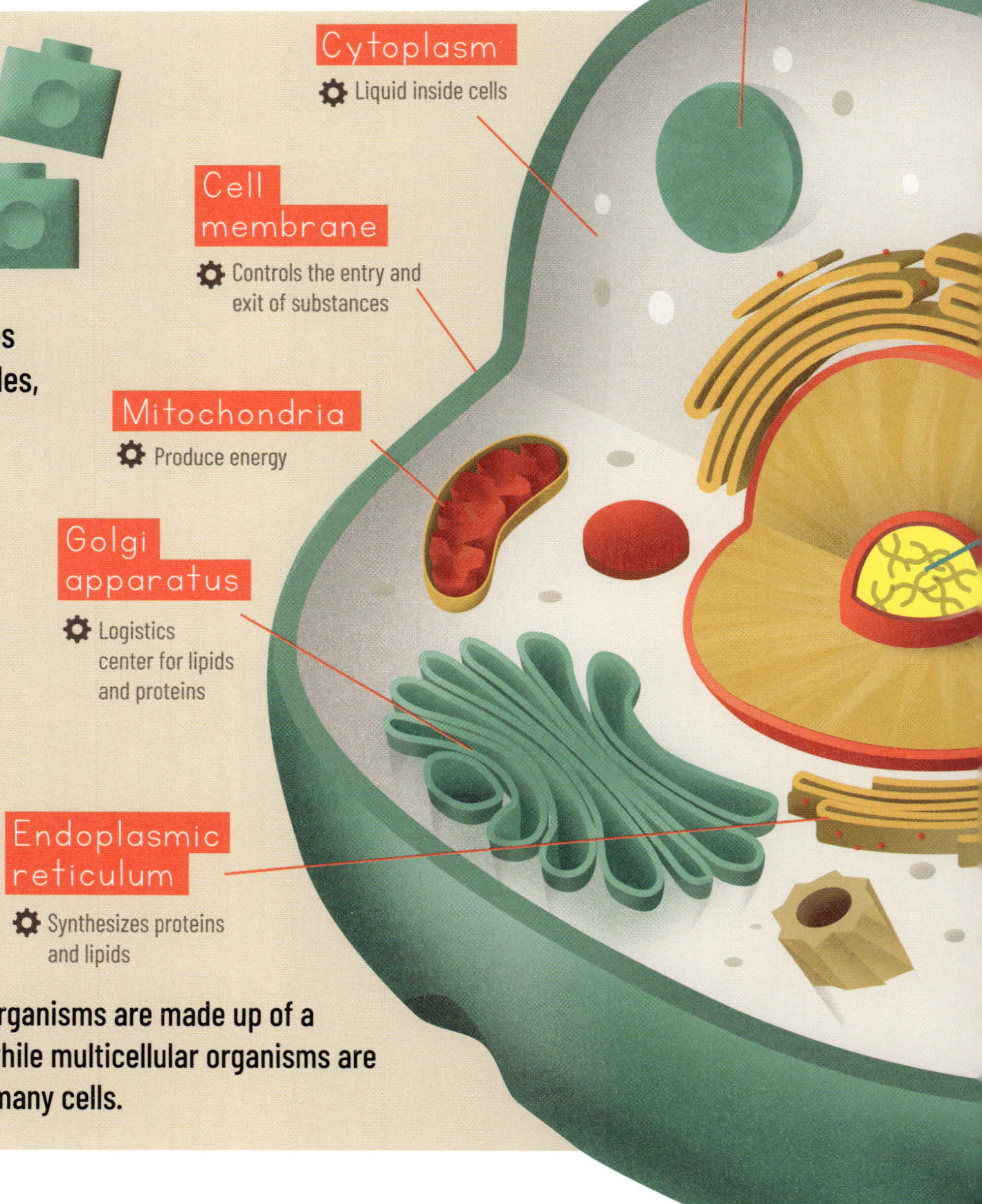

Lysosomes
Digest substances

Cytoplasm
Liquid inside cells

Cell membrane
Controls the entry and exit of substances

Mitochondria
Produce energy

Golgi apparatus
Logistics center for lipids and proteins

Endoplasmic reticulum
Synthesizes proteins and lipids

Unicellular organisms are made up of a single cell, while multicellular organisms are made up of many cells.

TYPES OF CELLS

Prokaryotic
Have no nucleus.
Unicellular beings.

Eukaryotic
Have a nucleus.
Can be unicellular or multicellular.

Ribosomes

Synthesize proteins

WHERE IS THE INFORMATION?

DNA (deoxyribonucleic acid) is the molecule that contains all the genetic and biological information of each living thing and is found in every one of its cells. To put it another way, it's an instruction manual.

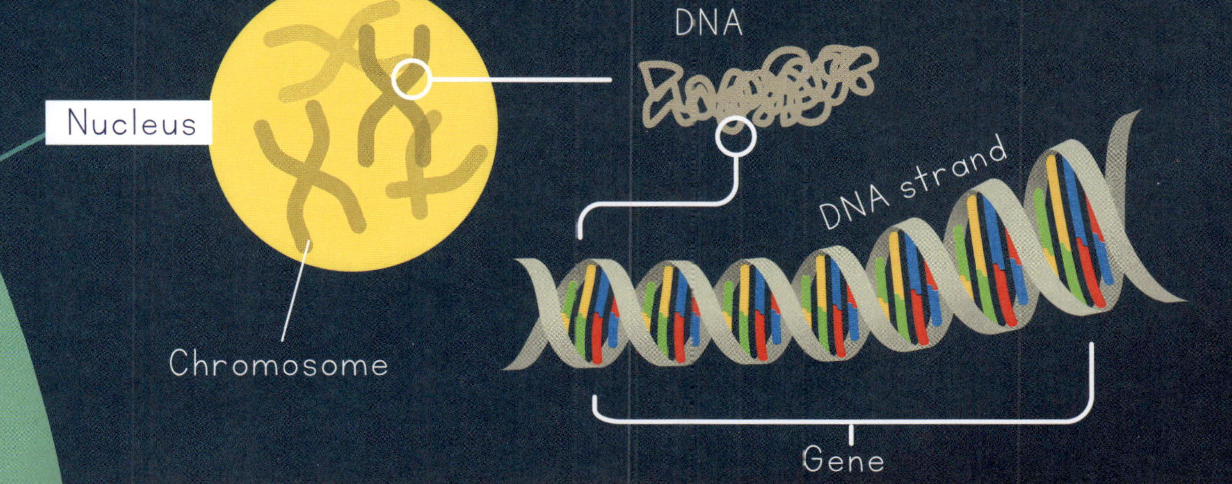

A gene is a bit or basic unit of hereditary information.
It is a DNA segment with the instructions that a cell needs to produce proteins, the molecules that make almost everything in a cell function.

All the cells in your body contain, in the DNA, the information about who or what you are (dog, microbe, etc.) and what you look like (tall, red hair, etc.). Each species has a certain number of genes. Humans have around 20,000 genes in our DNA. When organisms reproduce, they pass on their biological instructions, in other words, their DNA, to their offspring.

For example, every human baby is born from two cells, one from its mother (egg) and one from its father (sperm). Each of these cells contains a copy of different genes—the mother's genes and the father's genes—which combine to form a single distinct copy that the baby will inherit.

MUTATIONS

Mutations are changes in genetic information in the DNA of an organism.

Remember that the DNA of a living thing is like an instruction manual, affecting all aspects of its life: its behavior, appearance, and even, in the case of animals, its psychology.

WE ARE ALL MUTANTS!

We are very, very similar but different. If mutations didn't exist, individuals wouldn't have different traits; we would all be copies of our parents and there would be no evolution.

Causes of mutations

To allow us to grow or to repair damage to our bodies, cells split to create two new cells. This is called cell division.

Mutations can occur because the DNA doesn't copy well during cell division, or because it spontaneously breaks, or because external agents damage the DNA, such as exposure to radiation or viral infections.

Ideally, we don't want mutations because they can disrupt the proper function of cells and the organism. Therefore, cells have mechanisms to repair errors during DNA copying and stop mutations from happening.

Although DNA replication is a precise process, sometimes the cell doesn't manage to detect or repair the errors and a mutation occurs.

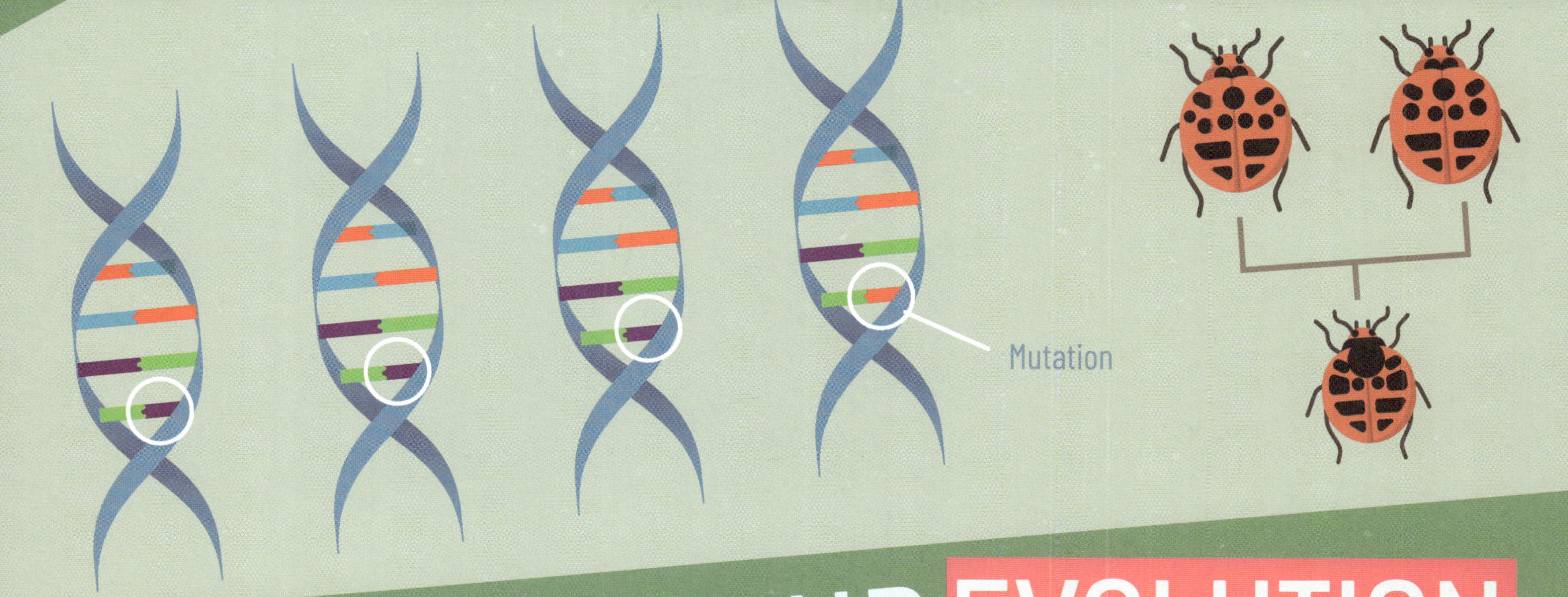

MUTATIONS AND EVOLUTION

MUTATIONS ARE VERY IMPORTANT FOR EVOLUTION AS THEY ARE THE MAIN SOURCE OF CHANGE IN LIVING THINGS.

For an organism, mutations can be:

Beneficial
They can help improve living things. Take bacteria, for example: if a mutation makes them resistant to an antibiotic, this gives them an adaptive advantage. However, this could be bad news for us because it makes it harder to treat infections.

Neutral
They have no effect.

Harmful
Some can cause diseases such as cancer.

Mutations are random, they just happen. A mutation isn't intended to improve or to harm anything. It just happens by chance and can have different consequences depending on genetics or the environment.

Imagine there's been a mutation in the hair color gene that makes it red. For this mutation to be inherited, it has to have occurred in the germ cells of the mother (egg) or the father (sperm). This mutation has led to the emergence of people with red hair. A new version of the gene has accidentally appeared in the population, a new allele for red hair.

NOT ALL MUTATIONS ARE USEFUL FOR EVOLUTION.
All cells in an organism contain DNA, so mutations can occur in many parts of our body. Depending on where they occur, they can be passed on to our descendants and inherited or not.

Mutations that CANNOT be inherited (somatic)
These happen in cells that are not germ cells, such as skin cells, and are not passed on to offspring. These mutations only affect the person during their lifetime, for example a freckle that appears on a skin cell and is duplicated as the cells divide.

Mutations that CAN be inherited (germline)
These occur in the parents' germ cells that are responsible for reproduction. These mutations are passed on to offspring.
THEY ARE IMPORTANT FOR EVOLUTION AS THEY CAN BE PASSED ON TO DESCENDANTS AND ARE THE SOURCE OF CHANGE AND DIVERSITY IN POPULATIONS.

WE'RE NOT ALL THE SAME (LUCKILY)

Not all individuals in a population are the same. In fact, it's true to say they're all different.

If we look at any species, we'll see that all the individuals within it are the same but different: each individual is unique and has characteristics that make it unique.

All these differences, all this VARIABILITY within a species, is mainly due to MUTATIONS and the combination of genes that parents pass on to their children.

Page 20

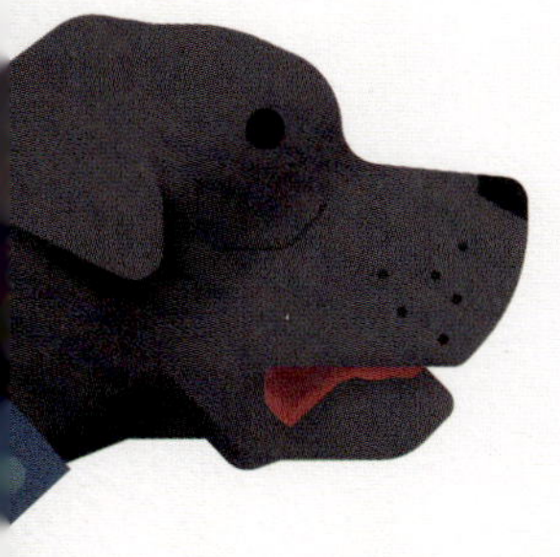

GENETIC VARIABILITY

But what makes us different? It is our individual genes that make us different from each other. The versions of a gene (alleles) we have for hair color, height, personality, tastes, muscle strength, and blood type largely determine what we are like.

And the fact we are all different is good for the species. The more variation there is in the individuals in a population, the better the species can adapt to a changing environment.

THE CONCEPT "GENETIC VARIABILITY" REFERS TO THE DIVERSITY AND FREQUENCY OF GENES IN A POPULATION.

Imagine we have two populations, each with 100 flowers.

If there are four different colors in a population of flowers, we can say that there are four versions of the color gene of the flower in that population. But, if we want to know how much genetic variability there is, we also need to look at the frequency—in other words, the number of flowers of each color in the population.

POPULATION 1

COLOR	INDIVIDUALS	ALLELE FREQUENCY
	25	**25 %**
	25	**25 %**
	25	**25 %**
	25	**25 %**

POPULATION 2

COLOR	INDIVIDUALS	ALLELE FREQUENCY
	5	**5 %**
	50	**50 %**
	25	**25 %**
	20	**20 %**

GENETIC VARIABILITY IS ESSENTIAL FOR A POPULATION TO SURVIVE, ADAPT, AND FLOURISH.

THE (MODERN) THEORY OF EVOLUTION

The modern theory of evolution is the scientific explanation that describes how species have changed over time compared to their ancestors.

All living species, including humans, share a common ancestor, the LUCA (see page 44), and have evolved from simpler life-forms over millions of years. Evolution is responsible for the similarities between living things, as well as the astonishing diversity of life. As we've already said, evolution happens when changes occur in a population that can be inherited by future generations or descendants.

We can think of evolution as changes that have happened over time in genes and their frequency in a population (the number of people that carry each allele).

But what causes these changes?

GENETIC DRIFT

This is the random change in the frequency of genes in a population. This can happen because of environmental factors or by chance in reproduction, and can lead to the loss of genetic variation in a population, especially a small one.

This is the movement of individuals from one population to another that causes genes to be exchanged. This can produce a new variation in a population or wipe out certain characteristics of another population.

MIGRATION

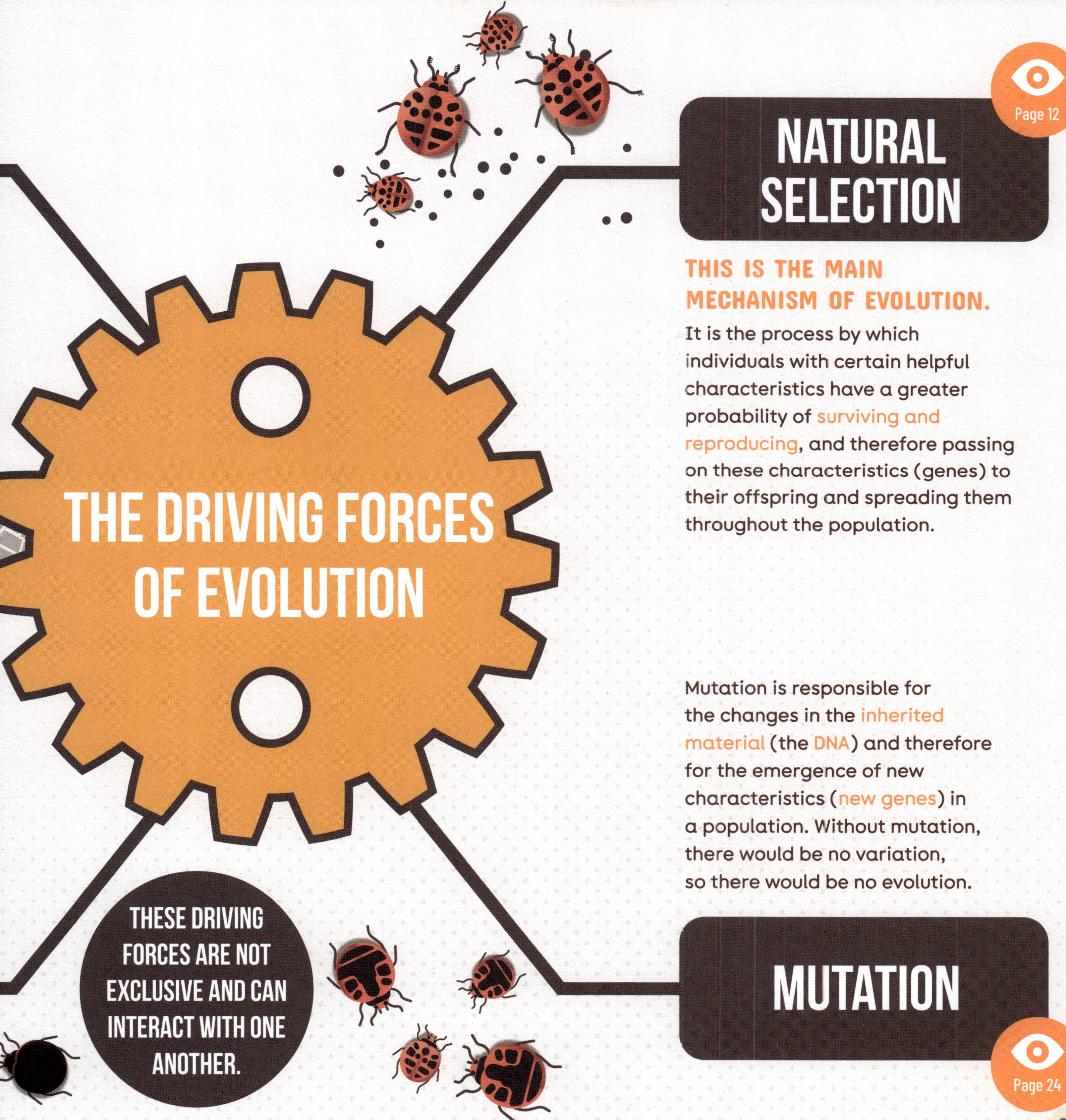

Page 12

NATURAL SELECTION

THIS IS THE MAIN MECHANISM OF EVOLUTION.

It is the process by which individuals with certain helpful characteristics have a greater probability of surviving and reproducing, and therefore passing on these characteristics (genes) to their offspring and spreading them throughout the population.

Mutation is responsible for the changes in the inherited material (the DNA) and therefore for the emergence of new characteristics (new genes) in a population. Without mutation, there would be no variation, so there would be no evolution.

MUTATION

Page 24

THE EVOLUTION OF LIFE

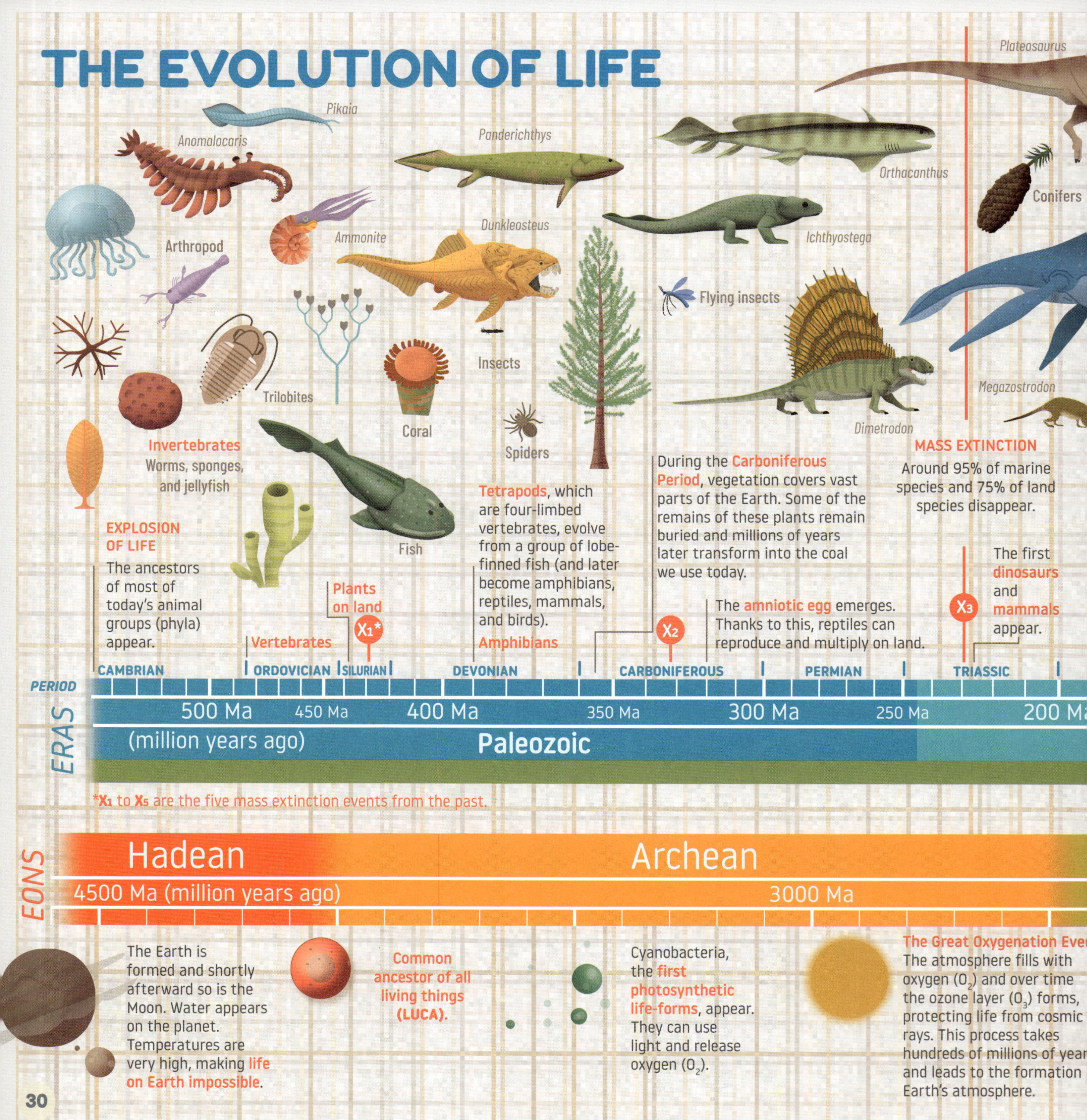

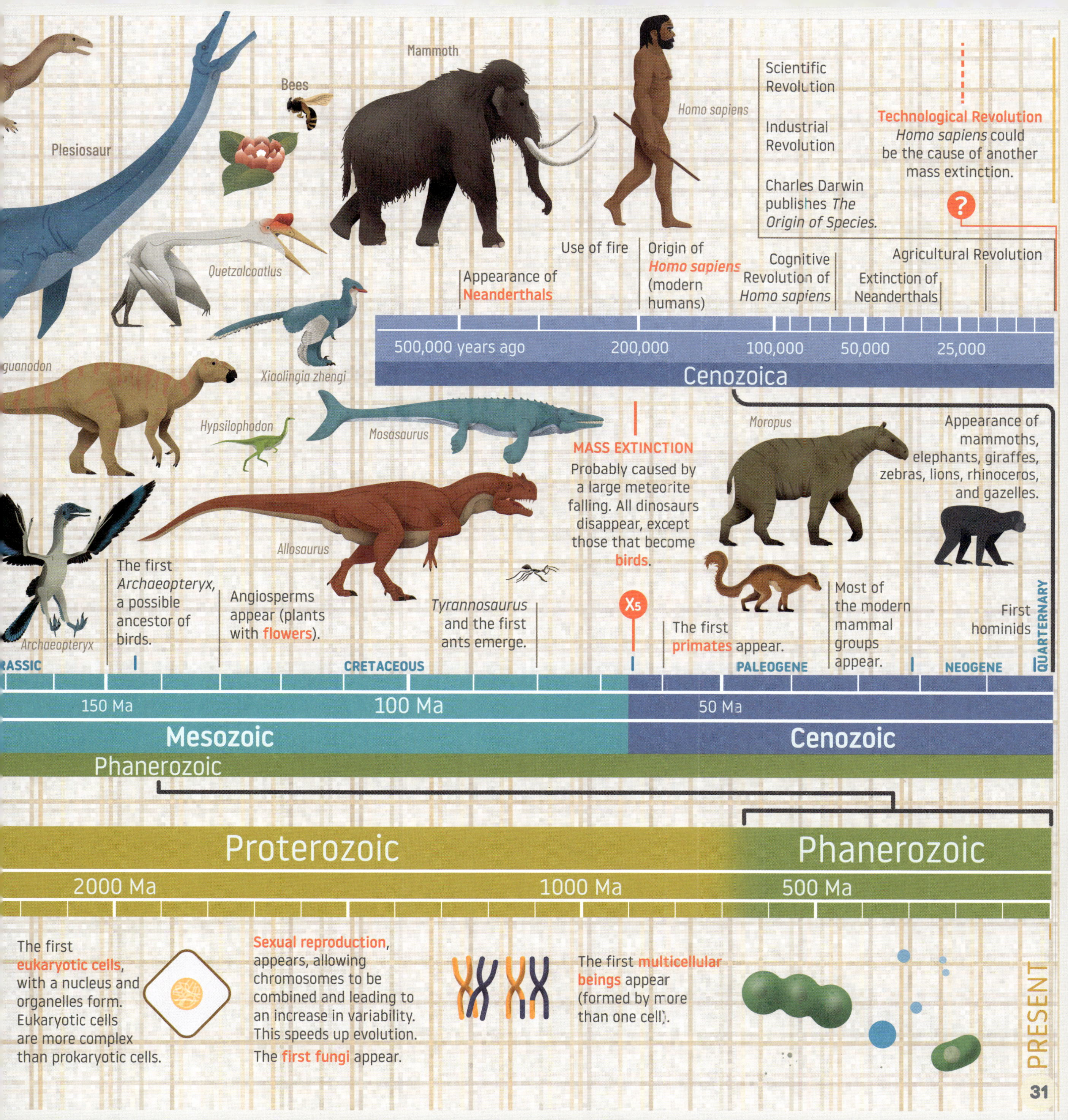

Plesiosaur
Bees
Mammoth
Homo sapiens
Scientific Revolution
Industrial Revolution
Charles Darwin publishes The Origin of Species.
Technological Revolution
Homo sapiens could be the cause of another mass extinction.
?
Quetzalcoatlus
Use of fire
Origin of Homo sapiens (modern humans)
Cognitive Revolution of Homo sapiens
Agricultural Revolution
Extinction of Neanderthals
Appearance of Neanderthals
500,000 years ago
200,000
100,000
50,000
25,000
Cenozoica
Iguanodon
Xiaolingia zhengi
Hypsilophodon
Mosasaurus
MASS EXTINCTION
Probably caused by a large meteorite falling. All dinosaurs disappear, except those that become birds.
Moropus
Appearance of mammoths, elephants, giraffes, zebras, lions, rhinoceros, and gazelles.
Allosaurus
The first Archaeopteryx, a possible ancestor of birds.
Angiosperms appear (plants with flowers).
Tyrannosaurus and the first ants emerge.
X5
The first primates appear.
Most of the modern mammal groups appear.
First hominids
Archaeopteryx
JURASSIC
CRETACEOUS
PALEOGENE
NEOGENE
QUARTERNARY
150 Ma
100 Ma
50 Ma
Mesozoic
Cenozoic
Phanerozoic
Proterozoic
Phanerozoic
2000 Ma
1000 Ma
500 Ma
The first eukaryotic cells, with a nucleus and organelles form. Eukaryotic cells are more complex than prokaryotic cells.
Sexual reproduction, appears, allowing chromosomes to be combined and leading to an increase in variability. This speeds up evolution.
The first fungi appear.
The first multicellular beings appear (formed by more than one cell).
PRESENT

SPECIATION

Speciation happens when a population divides into groups that for some reason remain separate in different regions or territories. Over time, the individuals in the different groups adapt to their environment and evolve in different directions. Eventually, they reach a point where they are so different that they cannot reproduce with each other. In other words, the different populations cannot exchange genes and each starts to become a different species from the original one.

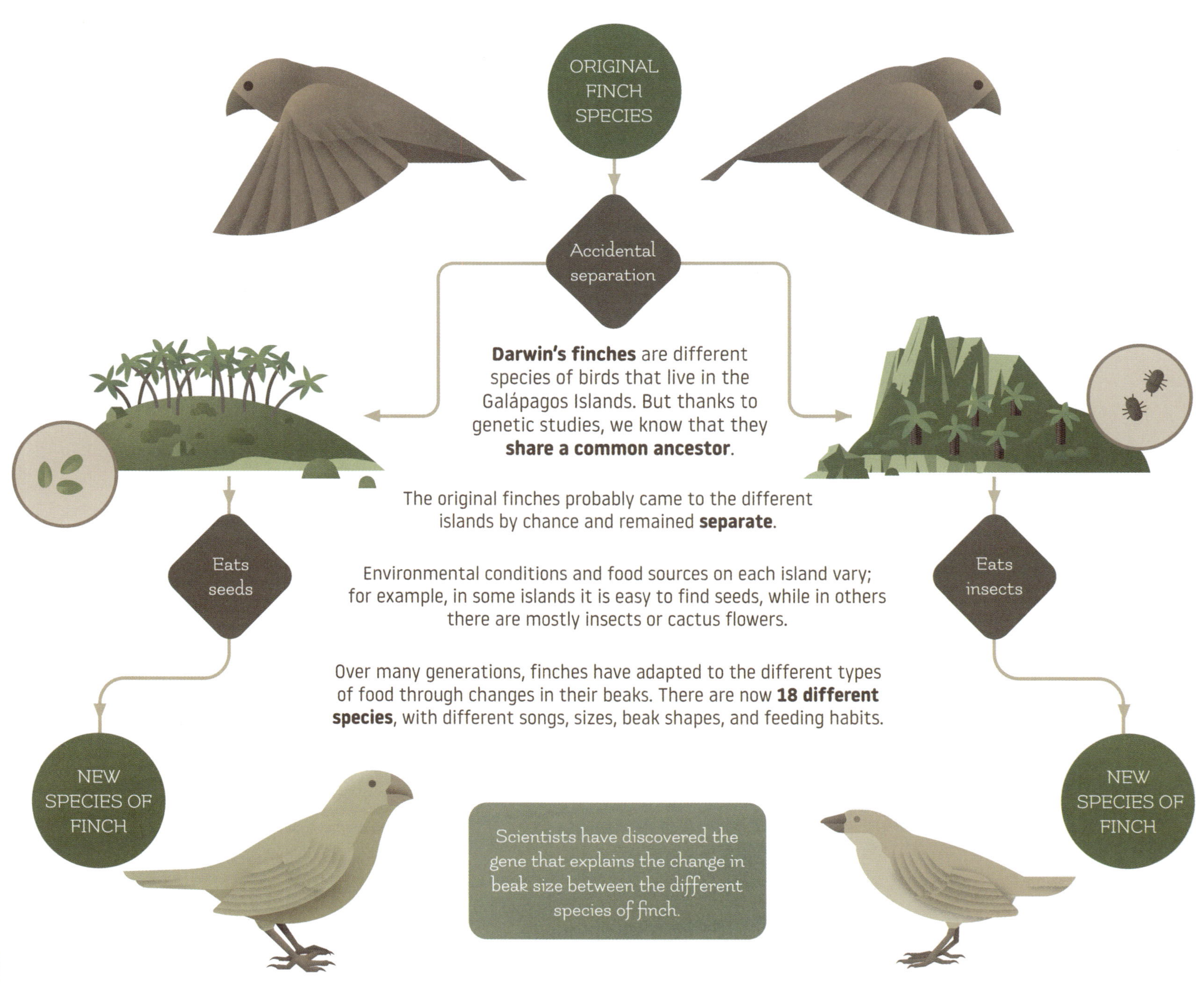

Darwin's finches are different species of birds that live in the Galápagos Islands. But thanks to genetic studies, we know that they **share a common ancestor**.

The original finches probably came to the different islands by chance and remained **separate**.

Environmental conditions and food sources on each island vary; for example, in some islands it is easy to find seeds, while in others there are mostly insects or cactus flowers.

Over many generations, finches have adapted to the different types of food through changes in their beaks. There are now **18 different species**, with different songs, sizes, beak shapes, and feeding habits.

Scientists have discovered the gene that explains the change in beak size between the different species of finch.

Microevolution and macroevolution

MICROEVOLUTION refers to the evolutionary changes that happen to a species in a particular area (a population) over relatively short timescales, such as generations or decades.

These changes are related to genetic variation within a population, and can lead to the population adapting to a changing environment and forming new inherited characteristics.

Microevolution is evolution on a small scale. We're only looking at a single branch of the tree of life.

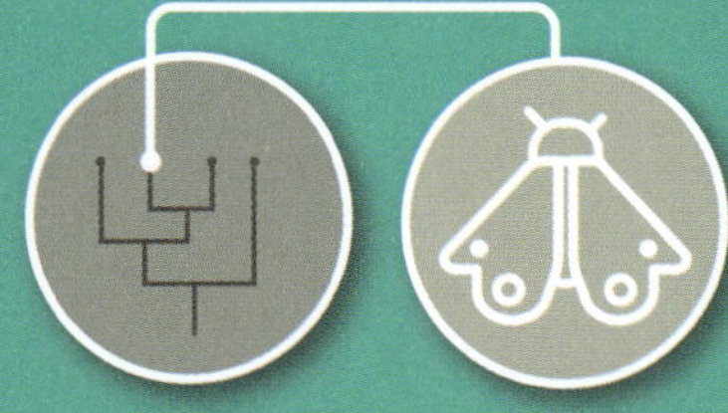

So, microevolution can lead to changes in the frequency of particular alleles in a population, which in turn can lead to adaptation to a changing environment. It can also lead to the appearance of new genetic variations, such as antibiotic resistance to bacteria or new color characteristics in an animal population.

SEE AN EXAMPLE OF MICROEVOLUTION ON PAGE 34.

MACROEVOLUTION refers to the evolutionary changes that occur on a large scale, such as new species being formed, groups of organisms diversifying, or complex characteristics evolving.

Macroevolution is evolution on a large scale. We look at a larger section of the tree of life to see its history.

Macroevolution involves evolutionary changes over long periods of time—millions of years. It leads to complex patterns of biodiversity, such as new species forming.

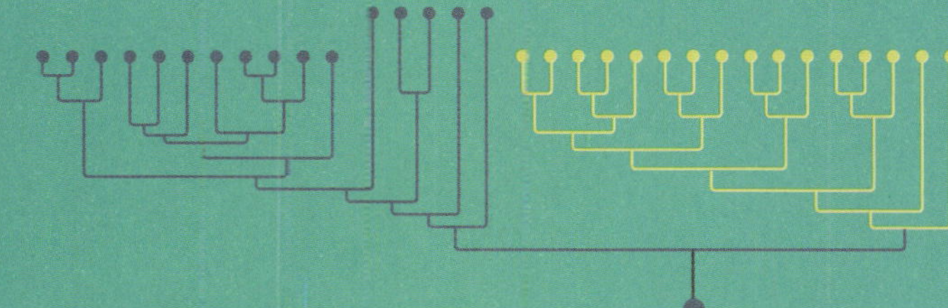

Macroevolution can also involve the evolution of complex characteristics such as new body structures like feathers and wings.

SEE AN EXAMPLE OF MACROEVOLUTION ON PAGE 36.

THE PEPPERED MOTH

THE PEPPERED MOTH (*Biston betularia*) USES THE TRUNK OF BIRCH TREES TO CAMOUFLAGE ITSELF FROM ITS PREDATORS, MAINLY BIRDS. THERE ARE TWO VARIETIES OF THIS MOTH:

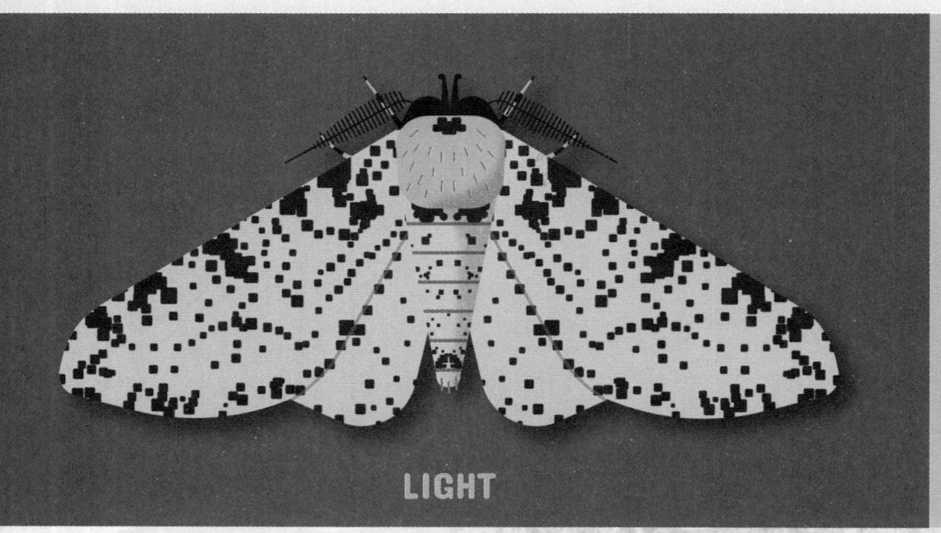

LIGHT

DARK

A lichen* grows on the bark of birch trees, giving them their characteristic appearance. The peppered moth is nocturnal, resting on the bark of birch trees during the day. The light variety of this moth remains completely camouflaged while resting on the tree trunk, unnoticed by its predators.

It is well adapted to pristine birch forests.

In the early 19th century, the dark variety of these moths was rare in England. This was because when they rested on the light-colored bark of birch trees, they were very visible. This made them easy prey for predators, unlike the light variety of moth, which went unnoticed.

*LICHEN is made up of two species, fungi and algae, which help each other to survive. The fungi get food from the algae and the algae provide the fungi with moisture so they don't dry out. This kind of association between species that is helpful to both is known as SYMBIOSIS.

During the Industrial Revolution, burning coal to produce steam caused a lot of pollution, creating soot that turned everything black, including birch trees. It also destroyed the lichen on the tree bark.

This meant the dark moth became the variety that was best adapted to its environment, as it remained camouflaged on the blackened bark, which helped it to survive and reproduce.

Meanwhile, the light peppered moths were highly visible and easy prey for birds. With this change in environment, the population of dark moths grew dramatically and the light moth population decreased until it almost disappeared.

In the mid-20th century, the English authorities realized that the pollution was very harmful to people and the environment and they banned the burning of so much coal. They managed to reduce the pollution and the birch forests returned to their original splendor: the trees became lighter and lichen began to grow on the bark again. The light moths were camouflaged and regained their evolutionary advantage over the dark moths, which once again became an easy target for predators.

The color of the moth population changed back to what it was before the Industrial Revolution. Today, the dark variety is rarely seen.

THIS IS A CLEAR EXAMPLE OF HOW A SPECIES EVOLVED THROUGH NATURAL SELECTION TO ADAPT TO ITS ENVIRONMENT.

THE EVOLUTION OF BIRDS

Birds are one of the most diverse and successful animal groups on Earth. They have evolved from a common ancestor that they shared with dinosaurs 150 million years ago.

The oldest known bird fossils date from the Upper Jurassic period, 150 million years ago. These fossils are quite similar to **theropods**, a group of carnivorous dinosaurs which includes the famous *Tyrannosaurus rex*, with which birds share a common ancestor. However, their closest relatives are the **dromaeosauridae**, a subgroup of theropods, commonly known as **raptors** (one of which is the *Velociraptor*). Raptors are much smaller, more agile, and more intelligent.

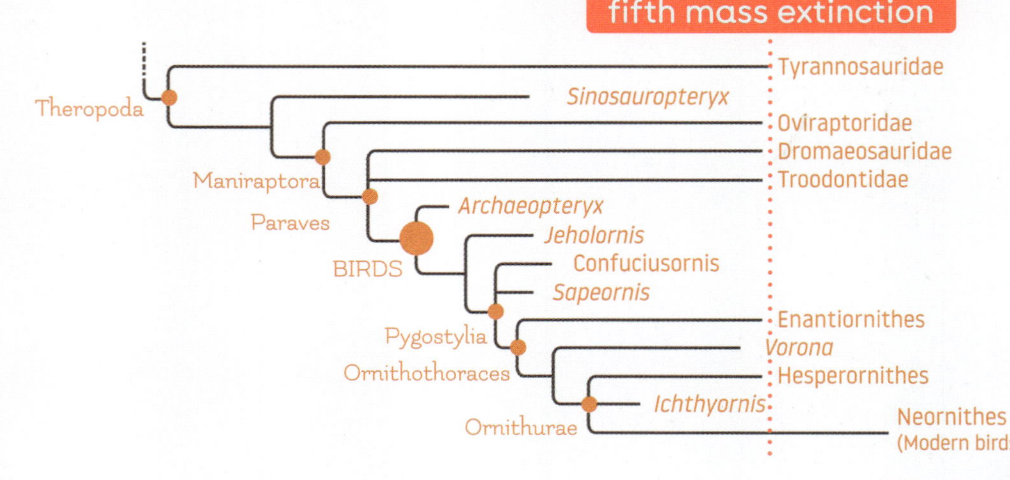

Velociraptor

Slow motion

How evolution creates new species has long intrigued scientists: does it happen suddenly over a few years, or more slowly and steadily over a very long time?

Scientists have been studying the evolution of birds for a long time, and thanks largely to the fossil record (see page 40) of birds and their ancestors, the dinosaurs, we know that the distinct traits of birds have built up gradually over millions of years. It seems that evolution on a large scale takes its time and moves in a leisurely fashion, in slow motion!

UPPER CRETACEOUS

Did feathers appear so animals could fly?

When we see an animal with feathers, we immediately think of a bird. But where did feathers come from? This question has puzzled scientists for a long time. Looking at clues we find in dinosaur fossils, we know that feathers didn't appear suddenly with the first birds but emerged much earlier with their reptile ancestors.

These first feathers were quite different from today's feathers. They were more like fluffy down, and scientists think that their main function was to keep the body warm. Fossils have shown that primitive feathers in dinosaurs then evolved to be more elaborate and colorful. We suspect they were used for display, like the male peacock, which opens its flashy tail feathers to impress females.

So, flight may have happened by accident. Dinosaurs with primitive wings discovered that they could glide when jumping from tree to tree, which gave them a certain advantage in the air. Over time, they developed larger pectoral muscles and longer arms in the form of wings until they eventually became modern birds.

Birds are a type of dinosaur, the only one to have survived to the present day; the others became extinct 66 million years ago.

Common loon

Thanks to these adaptations, birds are one of the most diverse and successful groups of animals. Today, there are 10,000 species of birds of all shapes and sizes, from the tiny hummingbird to the majestic eagle, as well as flightless birds such as the penguin, which can swim, and the large, fast ostrich.

Oviraptor

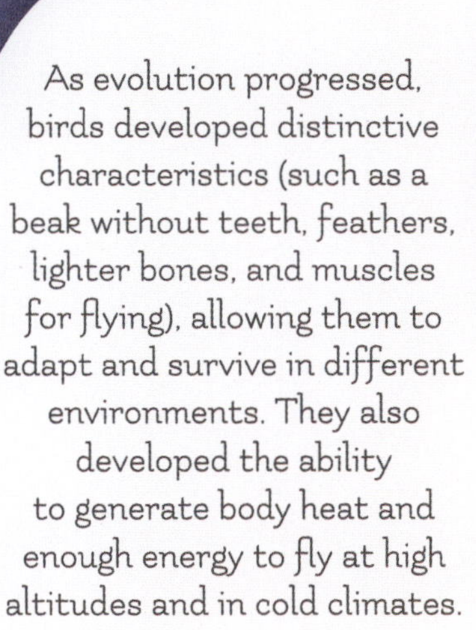

As evolution progressed, birds developed distinctive characteristics (such as a beak without teeth, feathers, lighter bones, and muscles for flying), allowing them to adapt and survive in different environments. They also developed the ability to generate body heat and enough energy to fly at high altitudes and in cold climates.

Emperor penguin

Chicken

Robin

PRESENT

COEVOLUTION

In nature, we can see examples of two or more species interacting and influencing each other's characteristics and adaptations. Such species evolve together. Biologists call this mutual evolution between species COEVOLUTION.

Plants with flowers provide nectar and pollen for insects such as bees, wasps, and butterflies. In return, the insects carry pollen from one flower to another to fertilize them. Flowers have adapted to communicate with insects through scent and color, so that the insects know where and how far away a flower is. Insects have adapted to gather food and pollen efficiently.

The bullhorn acacia has hollow thorns and produces nectar in its leaves. Some species of ants live in the thorns and feed on the nectar. In return, the ants protect the acacia by stinging any animals that try to eat the plant. This system is probably the result of coevolution. The plants developed hollow thorns and nectar to attract the ants, and the ants developed defense mechanisms to protect the plants. They influenced each other during evolution.

Pederson's cleaner shrimp provides a parasite cleaning service to various species of fish. When a fish approaches, the shrimp moves its antenna to show it is available to do the cleaning. Meanwhile, the fish sends signals to say that it's ready to be cleaned (and won't eat the shrimp!). The shrimp then eats the parasites on the fish's skin and cleans areas such as the inside of the gills or mouth. This way, the fish gets rid of the parasites and the shrimp gets some food. It's a win-win relationship.

ARTIFICIAL SELECTION

Artificial selection is a process whereby humans become involved in the evolution of certain species by selecting and breeding individuals with certain traits that we consider helpful or worthwhile. In doing so, we intervene in the heredity and genes of those species.

We often use artificial selection in **livestock** and **pet breeding** to produce varieties that are more productive, more resistant to disease, bigger, tastier, or have any other trait we consider useful.

Dogs are directly descended from wolves. They were domesticated thousands of years ago when humans interacted with wolves approaching settlements in search of food and shelter. It was a gradual evolutionary process in which humans selected the characteristics they were most interested in to **create different breeds of dogs**.

Over many generations, the selective breeding of these animals has led to a **huge variety** of sizes, shapes, and colors, and we have used dogs for hunting, sheep herding, protection, and companionship.

Thanks to a process called gene editing, we can change the DNA of living organisms and genetically modify some species for our benefit.

Over thousands of years, **farmers have selected seeds** from the most productive, tastiest, or best-looking crops to plant the next generation. **Over time**, this has led to **better harvests**.

So, we have big, tasty tomatoes, cows that produce lots of milk, hens that lay big eggs, horses that are very fast or very strong, watermelons with very few seeds, and rice that is resistant to disease and pests. We have even selected microorganisms, such as bacteria or fungi, to make foods such as bread and cheese.

FOSSILS

Fossils are the **remains of organisms** that existed in the past and are **preserved in rocks or other materials**. They can be the remains of whole organisms, such as skeletons, shells, teeth, feathers, leaves, and seeds, or they can be prints or tracks, such as footprints, worm tracks, or leaves imprinted in rocks.

These remains help us to **understand what the animals and plants** that lived hundreds, thousands, or even millions of years ago **were like** and how **life has evolved over time**.

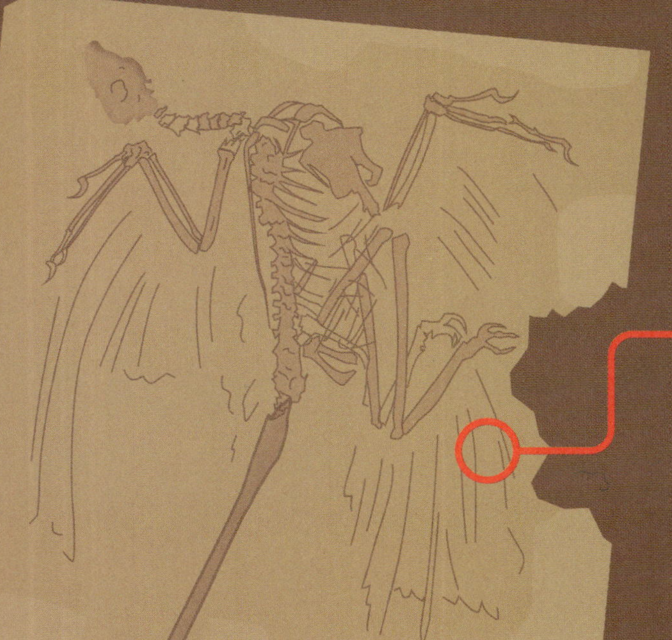

In this *Microraptor gui* fossil we can see fossilized feathers. It is one of the clues "written in the rock" that told scientists that winged dinosaurs once existed.

This and other similar fossils suggest that feathers evolved in dinosaurs a long time before birds appeared.

Through discoveries like this, we know that the ancestors of today's birds were dinosaurs.

Thanks to fossils, we know that *Microraptor gui* was a small dinosaur, around 3 feet (1 meter) long, that lived around 120 million years ago in what is now China.

We have no way of knowing what color it was or whether it could fly, although it could probably glide.

FOSSIL COMES FROM THE LATIN *FOSSILIS*, WHICH MEANS "DUG UP."

How are fossils formed?

1 Death and burial

When a living thing dies, it decomposes fast. To become a fossil, an animal or plant has to be buried very soon after its death so that decomposition slows down. Mud and sandy underwater beds are ideal for the formation of fossils.

2 Sedimentary layers

When animal or plant remains become buried, layers of rocks and minerals build up on top of them. Over time, these layers compact and compress until they become rock. The minerals in the remains of the organism are replaced by rock minerals and turn into a fossil.

3 Digging up

Paleontologists are scientists who study history through fossils. They are usually the ones who dig up fossilized remains found on a site. They have hammers, chisels, brushes, and other tools that they use very carefully to make sure they don't damage anything and preserve as much information as possible. Then they can study the fossils later in a laboratory and see what life was like in that place a long time ago.

There are also fossils in amber. They are usually insects and other small animals and plants that get trapped in tree resin. When the resin dries it forms amber, with the specimens perfectly preserved inside.

Another type of fossil is coprolite—fossilized excrement. This tells us what dinosaurs ate.

IMPROVABLE "DESIGNS"

The complexity we see in living things because of how individuals have adapted to their environment through evolution might lead us to think that nature has sculpted organisms that are perfectly designed to live in their habitats.

But natural selection is not a super-designer. It is more like a craftsperson who, through trial and error, has randomly crafted something, doing the best they can with what they have available—the best possible design for a particular environment, even if it's not perfect. Sometimes the design might even be strange or disadvantageous.

Humans have the same entrance for the airway (trachea) and for food and drink (esophagus) and this can cause us problems. For example, when we swallow something too quickly, the epiglottis (a flap in our throat) doesn't have time to close up our airway and food can get in and make us choke and cough. Other times food can get stuck in the throat and block the airway.

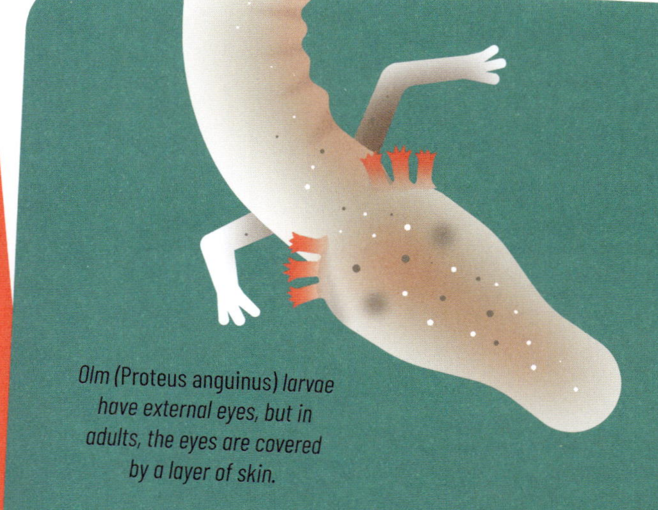

Olm (Proteus anguinus) larvae have external eyes, but in adults, the eyes are covered by a layer of skin.

Several species that live in caves or underground environments, such as some fish, insects, or amphibians, have lost the ability to see due to the lack of light in their habitat. The fact that these animals have eyes that do nothing could be considered bad design, as it means that energy is being used by an organ that has no adaptive advantage. It is a good example of evolution not being predesigned.

Remember that what we think of as "bad evolutionary design" is often because we're starting with what we know today thanks to science. Evolution doesn't always come up with the best solutions and it can leave behind the remains of structures that are either no longer needed or simply don't do what they're supposed to do very well.

IF SPECIES HAD BEEN DESIGNED TO BE PERFECT FROM THE START, 99% OF THEM WOULDN'T HAVE BECOME EXTINCT. >>>>

EXTINCTION

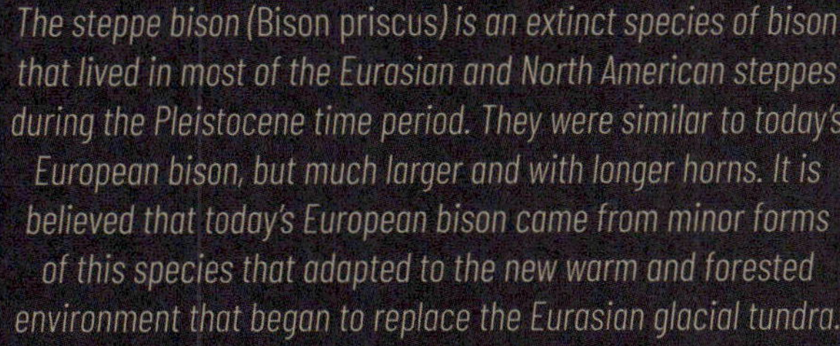

The steppe bison (Bison priscus) is an extinct species of bison that lived in most of the Eurasian and North American steppes during the Pleistocene time period. They were similar to today's European bison, but much larger and with longer horns. It is believed that today's European bison came from minor forms of this species that adapted to the new warm and forested environment that began to replace the Eurasian glacial tundra.

X

Human activity is the major cause of the biodiversity crisis on Earth.

We are probably on the verge of the sixth mass extinction.

Silphium was a plant used by the Greeks and Romans as a medicine and flavoring. It came from the Roman province of Cyrenaica (in what is now Libya) and was exported to the rest of the empire. It is thought to have become extinct in the 1st century.

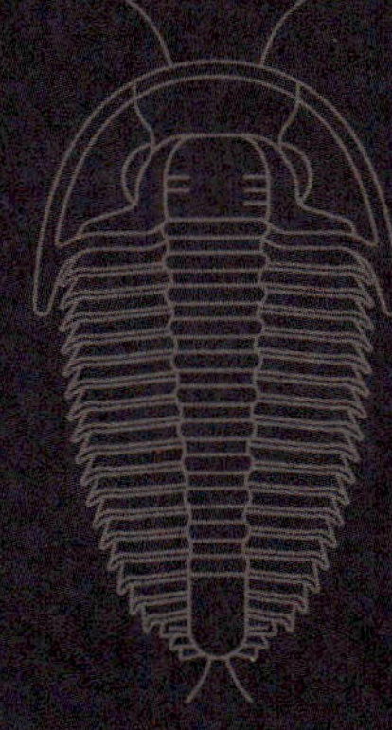

Trilobites were a class of arthropods, which appeared and diversified during the Cambrian period. Over 22,000 different species have been cataloged. After the Ordovician-Silurian mass extinction, only a few forms survived, and these disappeared during the Permian-Triassic mass extinction.

The baiji (Lipotes vexillifer) was a dolphin that lived in the Yangtze River in China. No specimens have been seen in the wild for forty years, so it is thought to be extinct, probably due to river pollution.

It has been calculated that 99% of the species that have lived on our planet are extinct.

This means that at some point, species that once lived on Earth were unable to adapt to a new environment. They weren't "designed" for the change in their habitat and didn't have time to evolve. Put simply, all the members of a species simply died, leaving no descendants.

MASS EXTINCTIONS We know from fossil records that there have been five periods in which most species disappeared due to catastrophic changes in the environment. Although mass extinctions were deadly events, they opened up the possibility for new life-forms to emerge and new species to flourish.

Percentage of species that disappeared:

85% X1 Ordovician-Silurian 444 million years ago (Ma)

82% X2 Devonian-Carboniferous 359 Ma

96% X3 Permian-Triassic 252 Ma

75% X4 Triassic-Jurassic 201 Ma

76% X5 Cretaceous-Paleogene 66 Ma

LUCA

(Last Universal Common Ancestor)

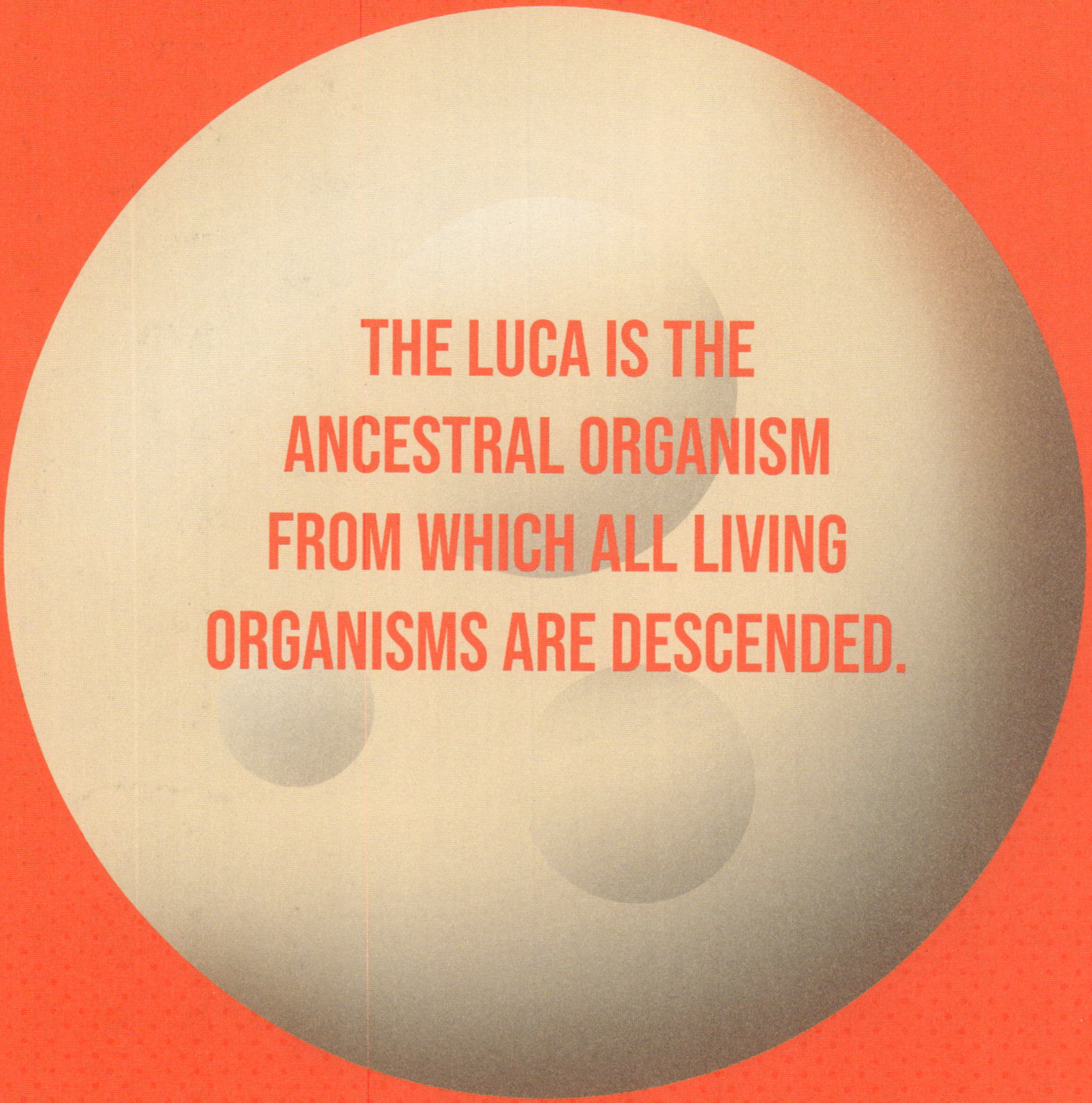

THE LUCA IS THE ANCESTRAL ORGANISM FROM WHICH ALL LIVING ORGANISMS ARE DESCENDED.

We can divide the diversity of life into three major groups: bacteria, archaea, and eukaryotes (which include humans and other animals, plants, and fungi). These three groups are very different from each other. Each is defined by its own very characteristic traits.

For example, eukaryotic cells are the only ones that have a cell nucleus where genetic material is stored. In contrast, archaea have a cell membrane that is very different from that of bacteria and eukaryotes.

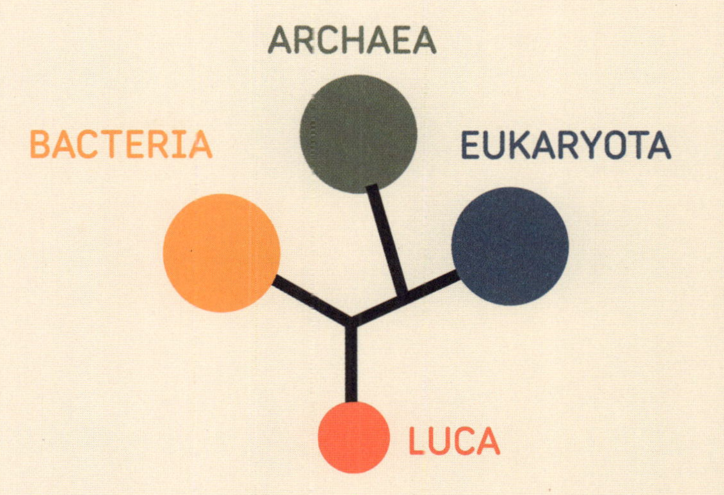

Despite our major differences, all living things on Earth share a number of characteristics that makes us think we have a common origin: the LUCA, a unicellular organism that lived around 4 billion years ago. From it, all the living things we know evolved.

IF THE GROUPS ARE SO DIFFERENT, HOW DO WE KNOW THAT EVERYTHING STARTED WITH A COMMON ANCESTOR?

Thanks to advances in bioinformatics (the way we capture and process large quantities of biological data), we have compared the proteins of around 1,000 of today's species.

From these studies, it has been estimated that at least 350 protein families found in species today have been inherited from a common ancestor: the LUCA. Based on this knowledge, we know that the LUCA already had a complex system of cellular functions.

There are still many questions to be answered about the LUCA, but with the development of new techniques and technologies, we will one day have a more complete and detailed view of the history of life.

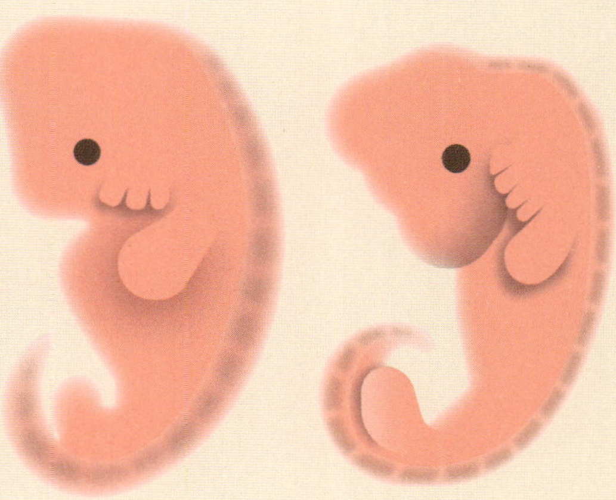

Human embryo   Bird embryo

The embryos of vertebrates (such as birds, mammals, and amphibians) are surprisingly similar in their first weeks of development. This was one of the factors that led scientists to believe we had common ancestors and a common evolutionary origin.

FUTURE HUMANS

LIKE ALL SPECIES, HUMANS ARE STILL EVOLVING.

We don't know what humans will be like in the future, but we will certainly continue to adapt to an environment that is heavily influenced by technology.

Our bodies will be affected by factors such as advances in medicine and genetics, nanotechnology implants, and living with artificial intelligence.

Also, if humanity becomes an interplanetary species, it will have to adapt to different gravity conditions and exposure to cosmic radiation.

We'll probably stop being *Homo sapiens* and become one or more different species. But this is a process that will take thousands and thousands of years.

Let's hope we don't become extinct before then and that we find a way to evolve into a better version of humanity!

EXTRATERRESTRIAL LIFE

WAS THE APPEARANCE OF LIFE ON EARTH SOMETHING RARE AND EXTRAORDINARY THAT ONLY HAPPENED THANKS TO THE EXCEPTIONAL CONDITIONS ON OUR PLANET, OR IS IT SOMETHING THAT CAN HAPPEN ON OTHER WORLDS?

The chemical elements that make up life (or at least life as we know it) are present in many parts of the universe, but until now the only planet we know of that is organized in such a highly complex way to support life is ours.

We do not know if there is life beyond Earth. For now, all we can do is search for traces of life (past or present) in environments similar to ours, on exoplanets* that have the right conditions—temperature, water, chemical elements—for life to develop.

*An exoplanet is a planet that orbits around a star other than the Sun; it belongs to a different planetary system from our solar system.

Thanks to the powerful telescopes we now have, thousands of exoplanets have been discovered. The strange thing is that these recently discovered distant worlds are nothing or very little like Earth. It won't be easy to detect life on a planet orbiting around a distant star but thanks to technological advances, we can search for remote signs of life.

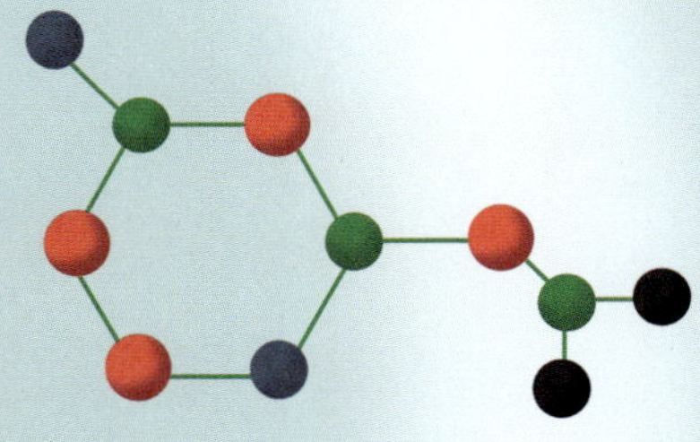

COULD THERE BE MORE LIFE IN THE UNIVERSE?

Complex organic molecules have been discovered in other planetary systems, suggesting that life could be quite common in the universe.

We know that there are billions of stars with their own exoplanets, so it seems quite probable that there is life on many of them. But perhaps life is something extremely rare and the chances of it existing in another part of the universe are minuscule.

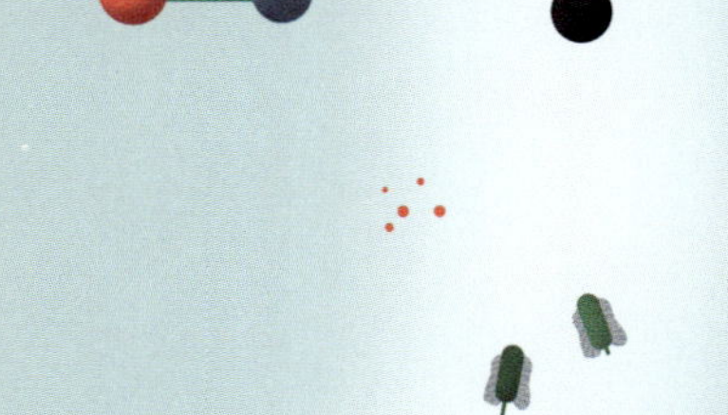

BUT WHAT ARE WE ACTUALLY LOOKING FOR?

As the only life we know of is life on Earth, we're trying to find something similar in other parts of the universe, such as plants, animals, or more likely, microorganisms similar to bacteria, archaea, or fungi that can live in harsh environments.

But perhaps the life we are looking for is very different from the one we know and isn't based on DNA, RNA, proteins, or other similar organic molecules. So how do we know we've found something if we don't know what it is and what it's like?

IT IS ONE THING TO FIND LIFE AND QUITE ANOTHER TO FIND LIFE-FORMS CAPABLE OF CREATING A CIVILIZATION. HUMAN CIVILIZATION HAS ONLY EXISTED FOR THE BLINK OF AN EYE IN THE HISTORY OF THE PLANET AND WE DON'T KNOW HOW LONG IT WILL LAST.

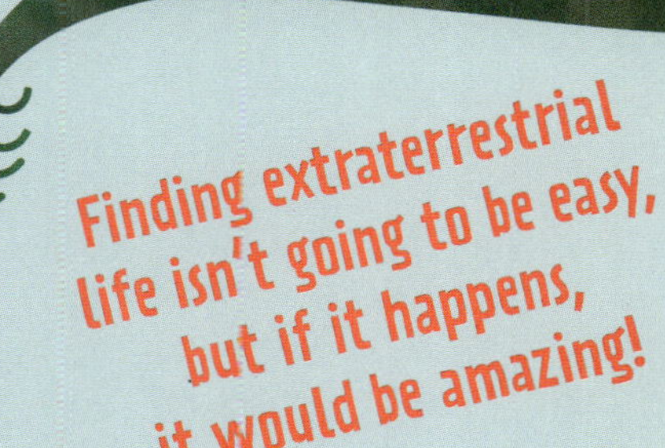

Finding extraterrestrial life isn't going to be easy, but if it happens, it would be amazing!

ACKNOWLEDGMENTS

Sheddad (@SheddadKF)—To **Sergi Pla Rabés** for proofreading the book. To **Helena** for proofreading and correcting the text and, most of all, for being there. And for **Tarek**, **Unai**, and **Inma**, of course. I love you all.

Eduard (@eduardaltarriba)—A thousand thanks to the people who have made this book possible, especially **Meli**, for her work and constant support in books as in life. And of course to **Per**, **Lourdes**, and **Ariadna**, for always being there.

And to all the scientists who, with their work, have made, are making, and will make it possible to reach further and further.

First published 2024 in English by Button Books, an imprint of Guild of Master Craftsman Publications Ltd, Castle Place, 166 High Street, Lewes, East Sussex, BN7 1XU, UK. © Sheddad Kaid-Salah Ferrón and Eduard Altarriba, 2023. English text © GMC Publications Ltd, 2024. ISBN: 978-1-78708-130-7. Distributed by Publishers Group West in the United States. All rights reserved. This translation of *My First Book of Evolution* is published by arrangement with Asterisc Agents. The right of Sheddad Kaid-Salah Ferrón and Eduard Altarriba to be identified as the authors of this work has been asserted in accordance with the Copyright, Designs and Patents Act 1988, sections 77 and 78. No part of this publication may be reproduced, stored in a retrieval system, or transmitted in any form or by any means without the prior permission of the publisher and copyright owner. While every effort has been made to obtain permission from the copyright holders for all material used in this book, the publishers will be pleased to hear from anyone who has not been appropriately acknowledged and to make the correction in future reprints. The publishers and authors can accept no legal responsibility for any consequences arising from the application of information, advice, or instructions given in this publication. A catalog record for this book is available from the British Library. For Alababalà, Design and Layout: Alababalà. Proofreading: Leticia Oyola. For GMC Publications, Publisher: Jonathan Bailey. Production Director: Jim Bulley. Senior Project Editor: Tom Kitch. Design Manager: Robin Shields. Copy Editor: Claire Saunders. Technical Consultant: Tiffany Taylor. English Translation: Andrea Reece. Color origination by GMC Reprographics. Printed and bound in China.